Table of Contents

Course Coordinator

CAD

EXPERIMENT 1

SOLID MODELING AND ASSEMBLY OF FLANGE COUPLING

Name: _____

Register Number: _____

Report Submitted on: _____

Important Instructions:

1. Lab reports must be submitted on the next working day of the lab class. Late submissions will entitle reduction in marks allotted for reports.

2. Plots must be glued neatly on the space provided.

3. All observations and programs should be hand written. Avoid attaching printouts.

4. Individual reports must be submitted for all the lab experiments.

5. Plagiarism will be treated very seriously.

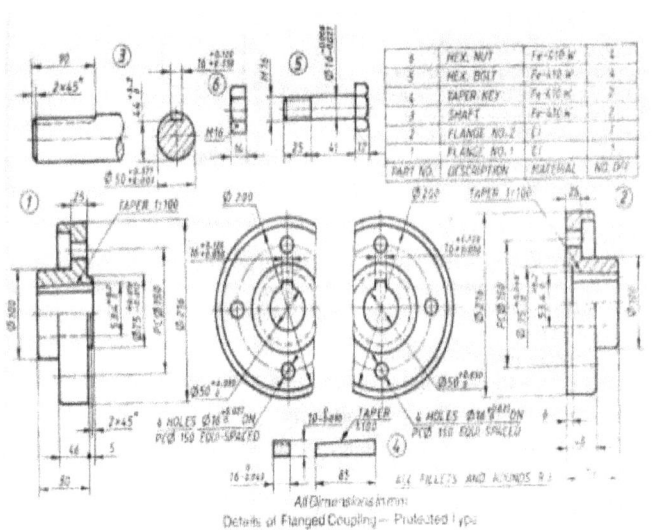

All Dimensions in mm
Details of Flanged Coupling – Protected Type

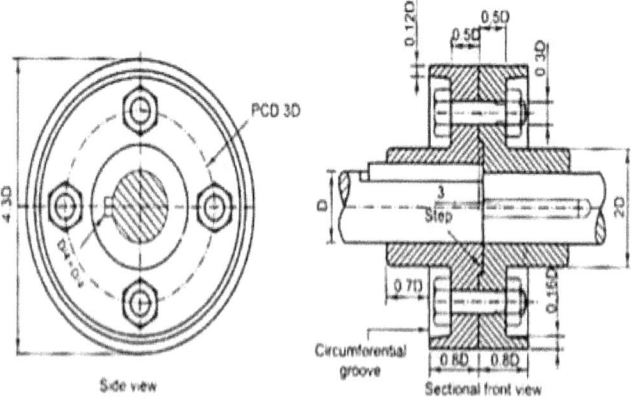

Side view

Sectional front view

Fig.1 Flange Coupling

ME 6611 CAD/CAM Laboratory
Department of Mechanical Engineering

SOLID MODELING AND ASSEMBLY OF FLANGE COUPLING

OBJECTIVE:

To draw the components of Flange Coupling and to assemble them using Solid works software.

SOFTWARE:

Solid Works 2008

EXPERIMENT THEORY:

A **coupling** is a device used to connect two shafts together at their ends for the purpose of transmitting power. Couplings do not normally allow disconnection of shafts during operation, however there are torque limiting couplings which can slip or disconnect when some torque limit is exceeded.

Flange coupling consists of two flanges keyed to the shafts. The flanges are connected together by means of bolts arranged on a circle concentric to shaft. Power is transmitted from driving shaft to flange on driving shaft through key, from flange on driving shaft to the flange on driven shaft through bolts and then to the driven shaft through key again. Projection is provided on one of the flanges and a corresponding recess is provided in the other for proper alignment. If in case failure of bolts occurs during the operation, the bolts may hit the operator in case of unprotected flange coupling. To avoid this, protective circumferential flanges are provided in the protected type flange coupling.

Flange of a protected type flange coupling has three distinct regions – inner hub, flanges and protective circumferential flanges. Following standard proportions are used in the design of flange coupling:

Outer diameter of hub,	$D = 2\,d$
Pitch circle diameter of bolts,	$D_1 = 3\,d$
Outer diameter of flange,	$D_2 = 4\,d$
Length of the hub,	$L = 1.5\,d$
Thickness of flange,	$t_f = 0.5\,d$
Thickness of protective circumferential flange,	$t_p = 0.25\,d$
where d is the diameter of shafts to be coupled.	

PROCEDURE:

1. Identify various parts to be created.
2. First enter into part environment and create the main part and create the main part of the assembly.
3. First identify whether the main part or the first to be created by protrusion or by revolution.
4. Select the sketch tool and then select the coincidental plane option and select any one of the standard 3 planes (i.e. front, right &top).
5. Create the cross-section profile as a closed one using the 2D commands available after completing the sketch, click open or return button and then click finish button.
6. For creating other parts, select sketch both parallel plane option or plane by 3points option and then select the required plane.
7. Construct the full cross section for portion and construct the half of the cross section and an axis line for revolution.
8. Do the protrusions by using protrusion command and the revolution by revolved protrusion command.
9. For constructing holes and cut-out, used hole command and cut-out command.
10. If we use hole command, change the diameter of the hole by using modify menu, resize hole option.
11. Use revolved cut-out command whenever needed.
12. Use the distance between option to maintain accurate distance between one edge and other edge or between one edge and to center the hole.
13. After constructing each part save it as a separate part file with extrusion* par.
14. Enter into assembly environment.
15. Assembly the various parts construct parts construct using the various assembly constrains available (planer, design, mate, axial align, connect etc).
16. After finishing assembly, check whether the various parts have been connected properly or not by rotating the view.
17. Save the assembly.

ASSEMBLY 3D DRAWING:

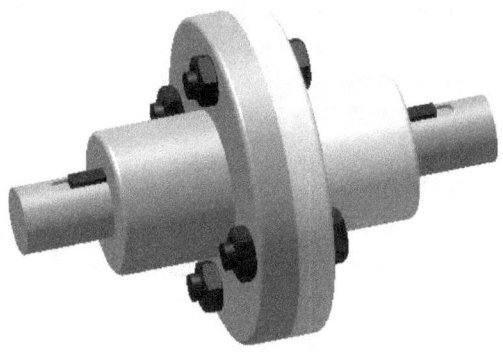

RESULTS & DISCUSSION:

_____

_____

_____

_____

VIVA QUESTIONS:

1. What is coupling?
2. What are the types of couplings?
3. What is protected type flange coupling?
4. What is use of protected type flange coupling?
5. What is the specification of protected type flange coupling?
6. What are the advantages of protected type flange coupling?

REPORT EVALUATION

S.NO	DESCRIPTION	WEIGHTAGE	MARK AWARDED
1	Part modeling	30	
2	Assembly & Drafting	30	
3	Results & Discussion	20	
4	Student Performance	10	
6	Viva-voce	10	
Total		100	

Signature of Course Coordinator

EXPERIMENT 2

SOLID MODELING AND ASSEMBLY OF PLUMMER BLOCK

Name: _____

Register Number: _____

Report Submitted on: _____

Important Instructions:

1. Lab reports must be submitted on the next working day of the lab class. Late submissions will entitle reduction in marks allotted for reports.

2. Plots must be glued neatly on the space provided.

3. All observations and programs should be hand written. Avoid attaching printouts.

4. Individual reports must be submitted for all the lab experiments.

5. Plagiarism will be treated very seriously.

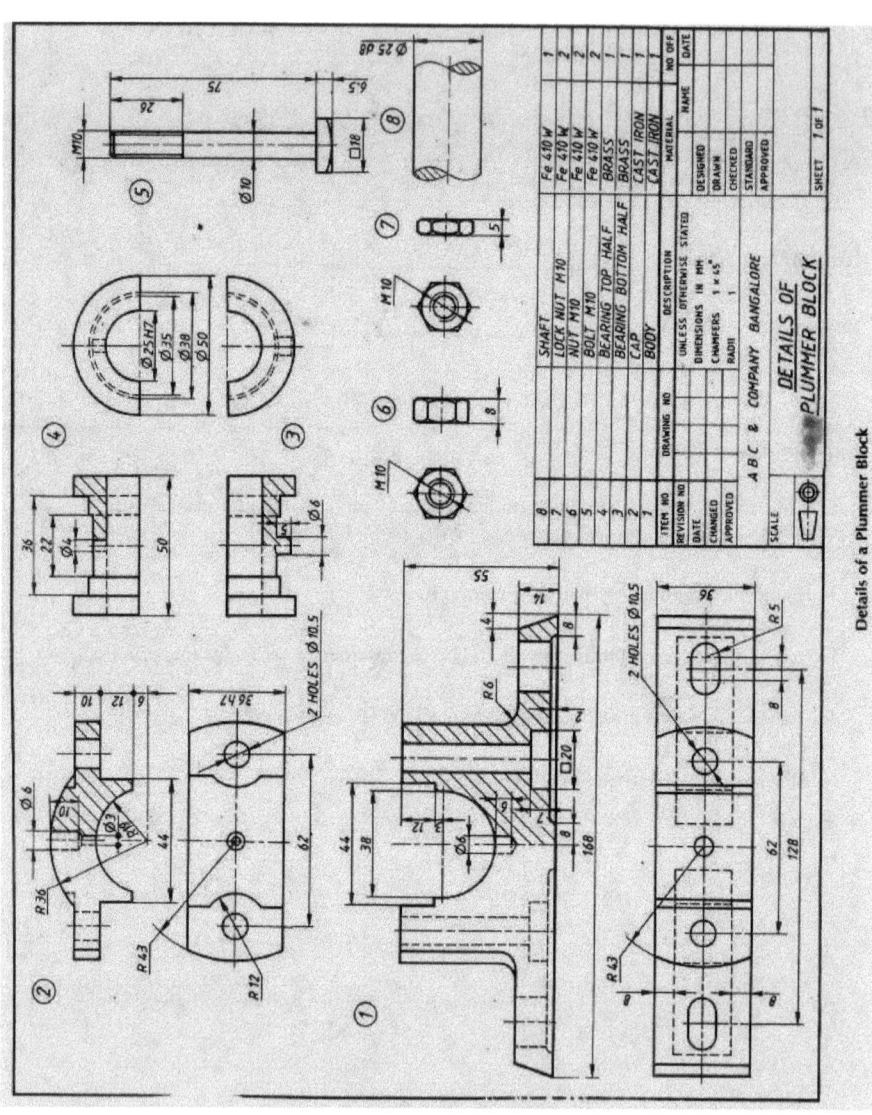

Fig 1. Plummer Block

ME 6611 CAD/CAM Laboratory
Department of Mechanical Engineering

SOLID MODELING AND ASSEMBLY OF PLUMMER BLOCK

OBJECTIVE:

To draw the components of Plummer block and to assemble them using Solid works software.

SOFTWARE:

Solid Works 2008.

DESCRIPTION ABOUT PLUMMER BLOCK:

A **pillow block**, also known as a **Plummer block** or **housed bearing unit**, is a pedestal used to provide support for a rotating shaft with the help of compatible bearings & various accessories. Housing material for a pillow block is typically made of cast iron or cast steel.

A pillow block usually refers to a housing with an included anti-friction bearing. A pillow block refers to any mounted bearing wherein the mounted shaft is in a parallel plane to the mounting surface, and perpendicular to the center line of the mounting holes, as contrasted with various types of flange blocks or flange units. A pillow block may contain a bearing with one of several types of rolling elements, including ball, cylindrical roller, spherical roller, tapered roller, or metallic or synthetic bushing. The type of rolling element defines the type of pillow block. These differ from "plumber blocks" which are bearing housings supplied without any bearings and are usually meant for higher load ratings and a separately installed bearing.

The fundamental application of both types is the same, which is to mount a bearing safely enabling its outer ring to be stationary while allowing rotation of the inner ring. The housing is bolted to a foundation through the holes in the base. Bearing housings may be either split type or solid type. Split type housings are usually two-piece housings where the cap and base may be detached, while others may be single-piece housings. Various sealing arrangements may be provided to prevent dust and other contaminants from entering the housing. Thus the housing provides a clean environment for the environmentally sensitive bearing to rotate free from contaminants while also retaining lubrication, either oil or grease, hence increasing its performance and duty cycle.

PROCEDURE:

1. Identify various parts to be created.
2. First enter into part environment and create the main part and create the main part of the assembly.
3. First identify whether the main part or the first to be created by protrusion or by revolution.
4. Select the sketch tool and then select the coincidental plane option and select any one of the standard 3 planes (i.e. front, right &top).
5. Create the cross-section profile as a closed one using the 2D commands available after completing the sketch, click open or return button and then click finish button.
6. For creating other parts, select sketch both parallel plane option or plane by 3points option and then select the required plane.
7. Construct the full cross section for portion and construct the half of the cross section and an axis line for revolution.
8. Do the protrusions by using protrusion command and the revolution by revolved protrusion command.
9. For constructing holes and cut-out, used hole command and cut-out command.
10. If we use hole command, change the diameter of the hole by using modify menu, resize hole option.
11. Use revolved cut out command whenever needed.
12. Use the distance between option to maintain accurate distance between one edge and other edge or between one edge and to centre the hole.
13. After constructing each part save it as a separate part file with extrusion* par.
14. Enter into assembly environment.
15. Assembly the various parts construct parts construct using the various assembly constrains available (planer, design, mate, axial align, connect etc.).
16. After finishing assembly, check whether the various parts have been connected properly or not by rotating the view.
17. Save the assembly.

ASSEMBLY 3D DRAWING:

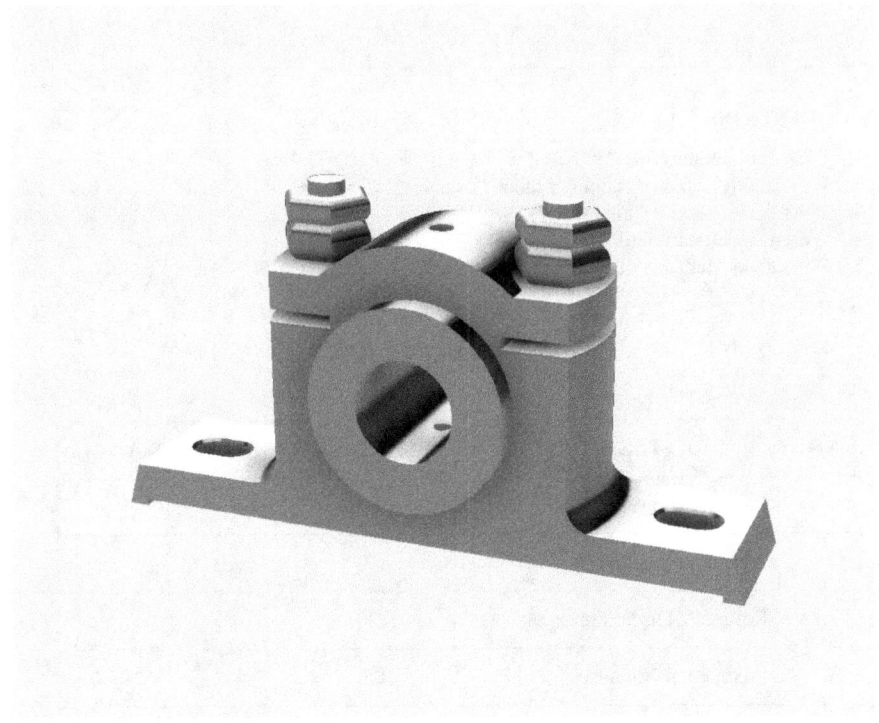

RESULTS & DISCUSSION:

_____

_____

_____

_____

_____

VIVA QUESTIONS:

1. What is Plummer block?
2. What are the types of Plummer block?
3. What is the use of Plummer block?
4. What is specification of Plummer block?
5. What are advantages of Plummer block?

REPORT EVALUATION

S.NO	DESCRIPTION	WEIGHTAGE	MARK AWARDED
1	Part modeling	30	
2	Assembly & Drafting	30	
3	Results & Discussion	20	
4	Student Performance	10	
6	Viva-voce	10	
	Total	100	

Signature of Course Coordinator

SOLID MODELING AND ASSEMBLY OF SCREW JACK

Name: _____

Register Number: _____

Report Submitted on: _____

Important Instructions:

1. Lab reports must be submitted on the next working day of the lab class. Late submissions will entitle reduction in marks allotted for reports.

2. Plots must be glued neatly on the space provided.

3. All observations and programs should be hand written. Avoid attaching printouts.

4. Individual reports must be submitted for all the lab experiments.

5. Plagiarism will be treated very seriously.

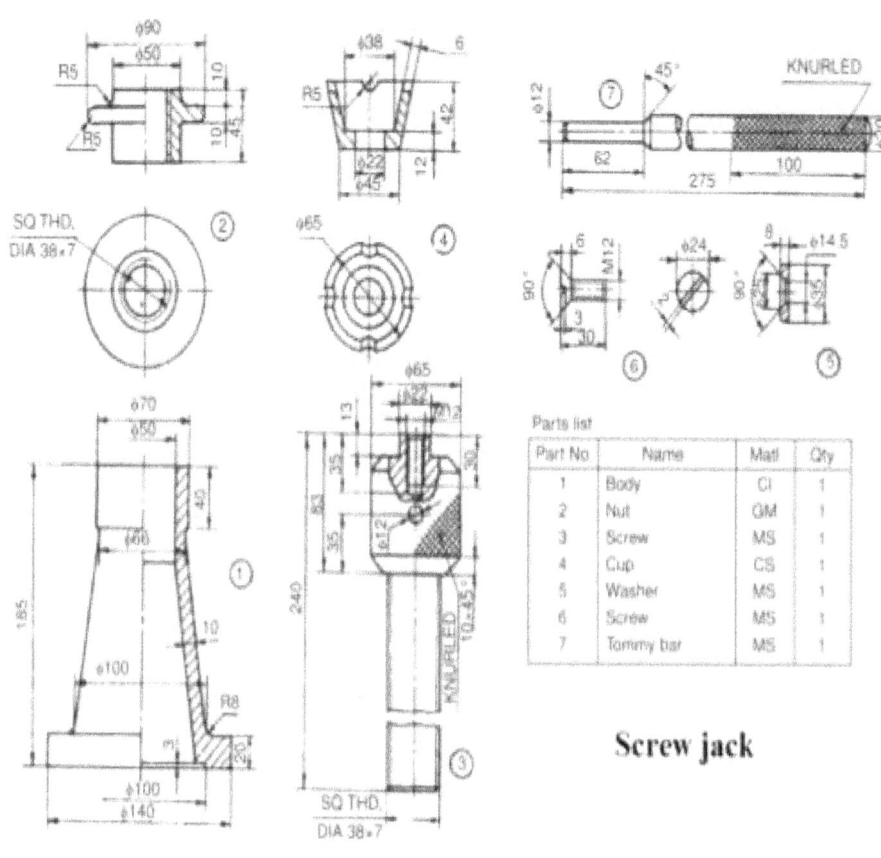

Fig 1. Screw Jack

SOLID MODELING AND ASSEMBLY OF SCREW JACK

OBJECTIVE:

To draw the components of screw jack and to assemble them using Solid works software.

SOFTWARE:

Solid Works 2008

DESCRIPTION ABOUT SCREW JACK:

A Screw Jack, manually operated is a contrivance to lift heavy object over a small height with a distinct Mechanical Advantages. It also serves as a supporting aid in the raised position. A screw Jack is actuated by a square threaded screw worked by applying a moderate effort at the end of a Tommy bar inserted into the hole of the head of the screw.

The body of the screw jack has an enlarged circular base which provides a large bearing area. A gun metal nut is tight fitted into the body at the top. A screw spindle is screwed through the nut. A load bearing cup is mounted at the top of the screw spindle and secured to it by a washer and a CSK screw. When the screw spindle is rotated, the load bearing cup moves only up or down along with the screw spindle but will not rotate with it. The Tommy bar is inserted into the hole in the head of the screw spindle only during working and will be detached when not in use.

PROCEDURE:

1. Identify various parts to be created.
2. First enter into part environment and create the main part and create the main part of the assembly.
3. First identify whether the main part or the first to be created by protrusion or by revolution.
4. Select the sketch tool and then select the coincidental plane option and select any one of the standard 3 planes (i.e. front, right &top).
5. Create the cross-section profile as a closed one using the 2D commands available after completing the sketch, click open or return button and then click finish button.
6. For creating other parts, select sketch both parallel plane option or plane by 3points option and then select the required plane.
7. Construct the full cross section for portion and construct the half of the cross section and an axis line for revolution.
8. Do the protrusions by using protrusion command and the revolution by revolved protrusion command.
9. For constructing holes and cut out, used hole command and cutout command.
10. If we use hole command, change the diameter of the hole by using modify menu, resize hole option.
11. Use revolved cut out command whenever needed.
12. Use the distance between option to maintain accurate distance between one edge and other edge or between one edge and to center the hole.
13. After constructing each part save it as a separate part file with extrusion* par.
14. Enter into assembly environment.
15. Assembly the various parts construct parts construct using the various assembly constrains available (planer, design, mate, axial align, connect etc).
16. After finishing assembly, check whether the various parts have been connected properly or not by rotating the view.
17. Save the assembly.

ASSEMBLY 3D DRAWING:

RESULTS & DISCUSSION:

_____

_____

_____

_____

_____

VIVA QUESTIONS:

1. What is screw jack?
2. What are the parts of screw jack?
3. What are the applications of screw jack?
4. What are advantages of screw jack?
5. What is the specification of screw jack?

REPORT EVALUATION

S.NO	DESCRIPTION	WEIGHTAGE	MARK AWARDED
1	Part modeling	30	
2	Assembly & Drafting	30	
3	Results & Discussion	20	
4	Student Performance	10	
6	Viva-voce	10	
Total		100	

Signature of Course Coordinator

EXPERIMENT 4

SOLID MODELING AND ASSEMBLY OF UNIVERSAL JOINT (COUPLING)

Name: _____

Register Number: _____

Report Submitted on: _____

Important Instructions:

1. Lab reports must be submitted on the next working day of the lab class. Late submissions will entitle reduction in marks allotted for reports.

2. Plots must be glued neatly on the space provided.

3. All observations and programs should be hand written. Avoid attaching printouts.

4. Individual reports must be submitted for all the lab experiments.

5. Plagiarism will be treated very seriously.

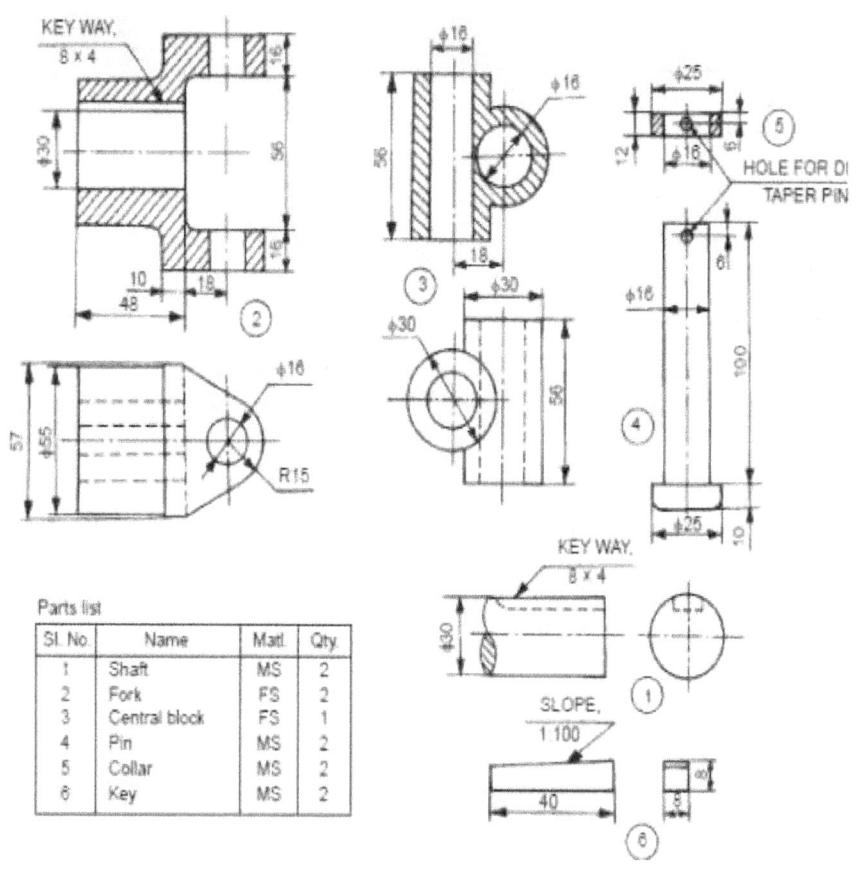

Parts list

Sl. No.	Name	Matl.	Qty.
1	Shaft	MS	2
2	Fork	FS	2
3	Central block	FS	1
4	Pin	MS	2
5	Collar	MS	2
6	Key	MS	2

Fig 1. Universal Coupling

SOLID MODELING AND ASSEMBLY OF UNIVERSAL JOINT (COUPLING)

OBJECTIVE:

To draw the components of universal joint and to assemble them using Solid works software.

SOFTWARE:

Solid Works 2008

DESCRIPTION ABOUT UNIVERSAL JOINT:

A **universal joint** (**universal coupling, U-joint, Cardan joint, Spicer** or **Hardy Spicer joint,** or **Hooke's joint**) is a joint or coupling in a rigid rod that allows the rod to "bend" in any direction, and is commonly used in shafts that transmit rotary motion. It consists of a pair of hinges located close together, oriented at 90° to each other, connected by a cross shaft. The universal joint is not a constant-velocity joint.

A universal joint is a positive, mechanical connection between rotating shafts, which are usually not parallel, but intersecting. They are used to transmit motion, power, or both.

The simplest and most common type is called the Cardan joint or Hooke joint. It consists of two yokes, one on each shaft, connected by a cross-shaped intermediate member called the spider. The angle between the two shafts is called the operating angle. It is generally, but n01 necessarily, constant during operation. Good design practice calls for low operating angles, often less than 25°, depending on the application. Independent of this guideline ne, mechanical interference in the construction of Cardan joints limits the operating angle to a maximum (often about 37½°), depending on its proportions.

Typical applications of universal joints include aircraft, appliances, control mechanisms, electronics, Instrumentation, medical and optical devices, ordnance, radio, sewing machines, textile machinery and tool drives.

Universal joints are available in steel or in thermoplastic body members. Universal joints made of steel have maximum load-carrying capacity for a given size. Universal joints with thermoplastic body members are used in light industrial applications in which their self-lubricating feature, light weight, negligible backlash, corrosion resistance and capability for high-speed operation are significant advantages.

Universal joints of special construction, such as ball-jointed universals are also available. These are used for high-speed operation and for carrying large torques. They are available both in miniature and standard sizes.

PROCEDURE:

1. Identify various parts to be created.
2. First enter into part environment and create the main part and create the main part of the assembly.
3. First identify whether the main part or the first to be created by protrusion or by revolution.
4. Select the sketch tool and then select the coincidental plane option and select any one of the standard 3 planes (i.e. front, right &top).
5. Create the cross-section profile as a closed one using the 2D commands available after completing the sketch, click open or return button and then click finish button.
6. For creating other parts, select sketch both parallel plane option or plane by 3points option and then select the required plane.
7. Construct the full cross section for portion and construct the half of the cross section and an axis line for revolution.
8. Do the protrusions by using protrusion command and the revolution by revolved protrusion command.
9. For constructing holes and cut-out, used hole command and cut-out command.
10. If we use hole command, change the diameter of the hole by using modify menu, resize hole option.
11. Use revolved cut-out command whenever needed.
12. Use the distance between option to maintain accurate distance between one edge and other edge or between one edge and to center the hole.
13. After constructing each part save it as a separate part file with extrusion* par.
14. Enter into assembly environment.
15. Assembly the various parts construct parts construct using the various assembly constrains available (planer, design, mate, axial align, connect etc.).
16. After finishing assembly, check whether the various parts have been connected properly or not by rotating the view.
17. Save the assembly.

ASSEMBLY 3D DRAWING:

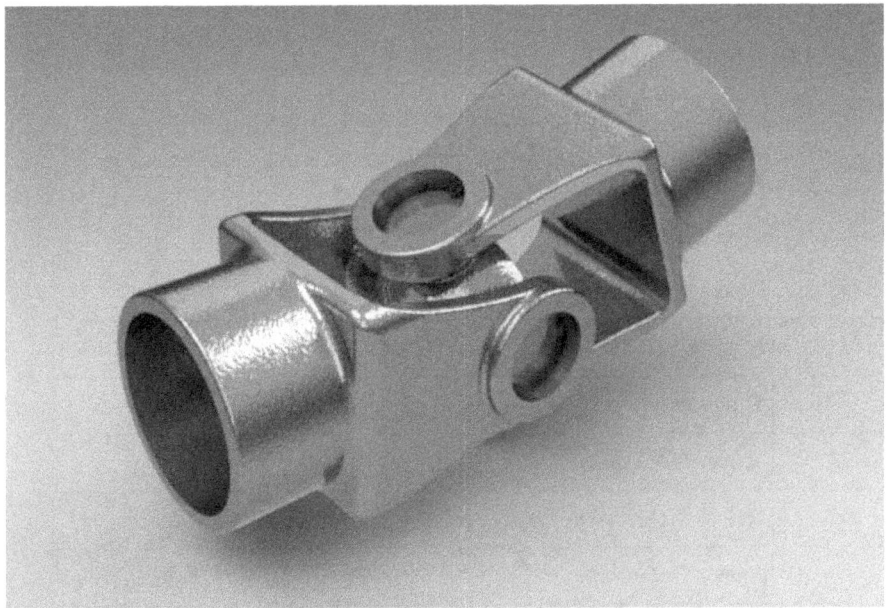

RESULTS & DISCUSSION:

_____

_____

_____

_____

_____

_____

VIVA QUESTIONS:

1. What is universal coupling?
2. What are the parts of universal coupling?
3. What are the applications of universal coupling?
4. What are advantages of universal coupling?

REPORT EVALUATION

S.NO	DESCRIPTION	WEIGHTAGE	MARK AWARDED
1	Part modeling	30	
2	Assembly & Drafting	30	
3	Results & Discussion	20	
4	Student Performance	10	
6	Viva-voce	10	
	Total	100	

Signature of Course Coordinator

EXPERIMENT 5

SOLID MODELING AND ASSEMBLY OF STUFFING BOX

Name: _____

Register Number: _____

Report Submitted on: _____

Important Instructions:

1. Lab reports must be submitted on the next working day of the lab class. Late submissions will entitle reduction in marks allotted for reports.

2. Plots must be glued neatly on the space provided.

3. All observations and programs should be hand written. Avoid attaching printouts.

4. Individual reports must be submitted for all the lab experiments.

5. Plagiarism will be treated very seriously.

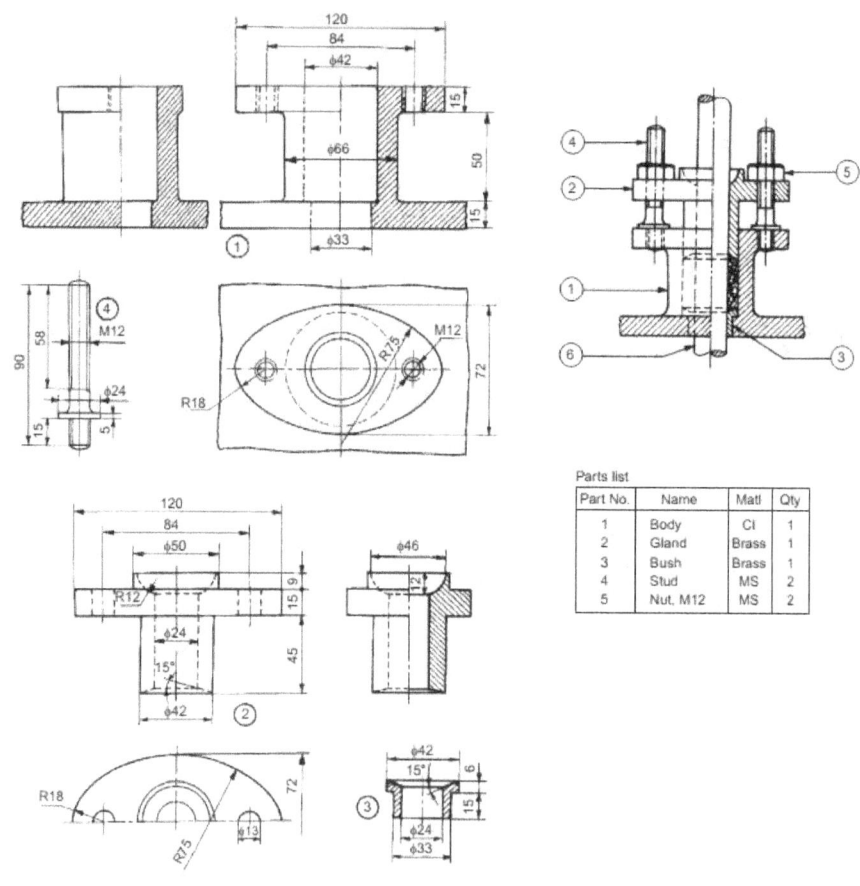

| Parts list | | | |
Part No.	Name	Matl	Qty
1	Body	CI	1
2	Gland	Brass	1
3	Bush	Brass	1
4	Stud	MS	2
5	Nut, M12	MS	2

Fig.1 Stuffing box

SOLID MODELING AND ASSEMBLY OF STUFFING BOX

OBJECTIVE:

To draw the components of stuffing box and to assemble them using Solid works software.

SOFTWARE:

Solid Works 2008.

DESCRIPTION ABOUT STUFFING BOX:

A **stuffing box** is an assembly which is used to house a gland seal. It is used to prevent leakage of fluid, such as water or steam, between sliding and turning parts of machine elements.

A stuffing box of a sailboat will have a stern tube that's slightly bigger than the prop shaft. It will also have packing nut threads or a gland nut. The packing is inside the gland nut and creates the seal. The shaft is wrapped by the packing and put in the gland nut. Through tightening it onto the stern tube, the packing is compressed, creating a seal against the shaft. Creating a proper plunger alignment is critical for correct flow and a long wear life. Stuffing box components are of stainless steel, brass or other application-specific materials.

A gland is a general type of stuffing box, used to seal a rotating or reciprocating shaft against a fluid. The most common example is in the head of a tap (faucet) where the gland is usually packed with string which has been soaked in tallow or similar grease. The gland nut allows the packing material to be compressed to form a watertight seal and prevent water leaking up the shaft when the tap is turned on. The gland at the rotating shaft of a centrifugal pump may be packed in a similar way and graphite grease used to accommodate continuous operation.

In a common type of stuffing box, rings of braided fiber, known as shaft packing or gland packing, form a seal between the shaft and the stuffing box. A traditional variety of shaft packing comprises a square cross-section rope made of flax or hemp impregnated with wax and lubricants. A turn of the adjusting nut compresses the shaft packing. Ideally, the compression is just enough to make the seal both watertight when the shaft is stationary and drip slightly when the shaft is turning. The drip rate must be at once sufficient to lubricate and cool the shaft and packing, but not so much as could sink an unattended boat.

PROCEDURE:

1. Identify various parts to be created.

2. First enter into part environment and create the main part and create the main part of the assembly.

3. First identify whether the main part or the first to be created by protrusion or by revolution.

4. Select the sketch tool and then select the coincidental plane option and select any one of the standard 3 planes (i.e. front, right &top).

5. Create the cross-section profile as a closed one using the 2D commands available after completing the sketch, click open or return button and then click finish button.

6. For creating other parts, select sketch both parallel plane option or plane by 3points option and then select the required plane.

7. Construct the full cross section for portion and construct the half of the cross section and an axis line for revolution.

8. Do the protrusions by using protrusion command and the revolution by revolved protrusion command.

9. For constructing holes and cut-out, used hole command and cut-out command.

10. If we use hole command, change the diameter of the hole by using modify menu, resize hole option.

11. Use revolved cut-out command whenever needed.

12. Use the distance between option to maintain accurate distance between one edge and other edge or between one edge and to center the hole.

13. After constructing each part save it as a separate part file with extrusion* par.

14. Enter into assembly environment.

15. Assembly the various parts construct parts construct using the various assembly constrains available (planer, design, mate, axial align, connect etc.).

16. After finishing assembly, check whether the various parts have been connected properly or not by rotating the view.

17. Save the assembly.

ASSEMBLY 3D DRAWING:

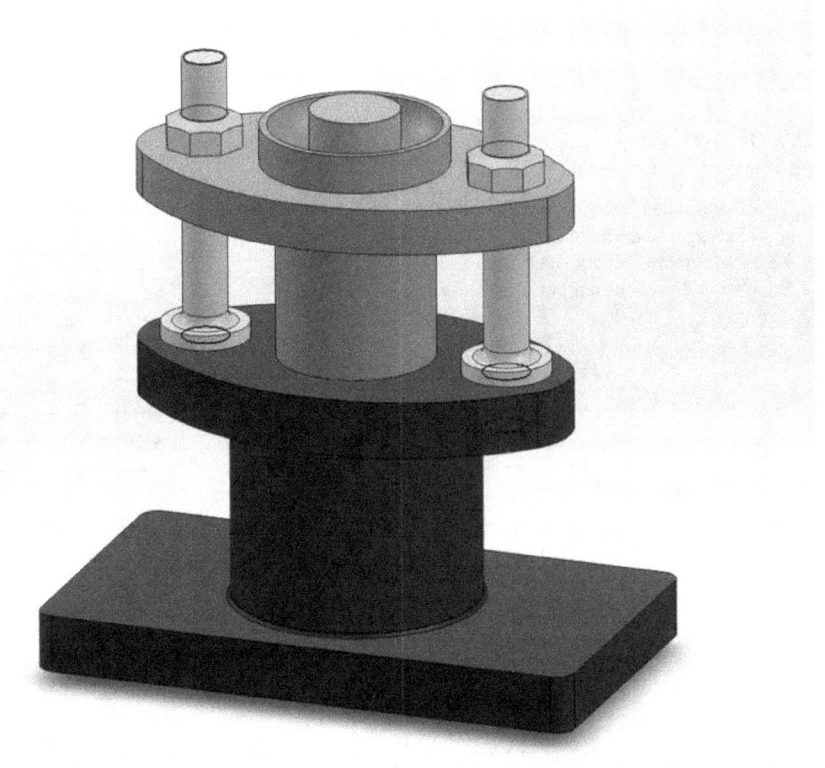

RESULTS & DISCUSSION:

_____

_____

_____

_____

_____

VIVA QUESTIONS:
1. What is stuffing box?
2. What are the parts of stuffing box?
3. What are the applications of stuffing box?
4. What are advantages of stuffing box?
5. What is the specification of stuffing box?

REPORT EVALUATION

S.NO	DESCRIPTION	WEIGHTAGE	MARK AWARDED
1	Part modeling	30	
2	Assembly & Drafting	30	
3	Results & Discussion	20	
4	Student Performance	10	
6	Viva-voce	10	
Total		100	

Signature of Course Coordinator

SOLID MODELING AND ASSEMBLY OF KNUCKLE JOINT

Name: _____

Register Number: _____

Report Submitted on: _____

Important Instructions:

1. Lab reports must be submitted on the next working day of the lab class. Late submissions will entitle reduction in marks allotted for reports.

2. Plots must be glued neatly on the space provided.

3. All observations and programs should be hand written. Avoid attaching printouts.

4. Individual reports must be submitted for all the lab experiments.

5. Plagiarism will be treated very seriously.

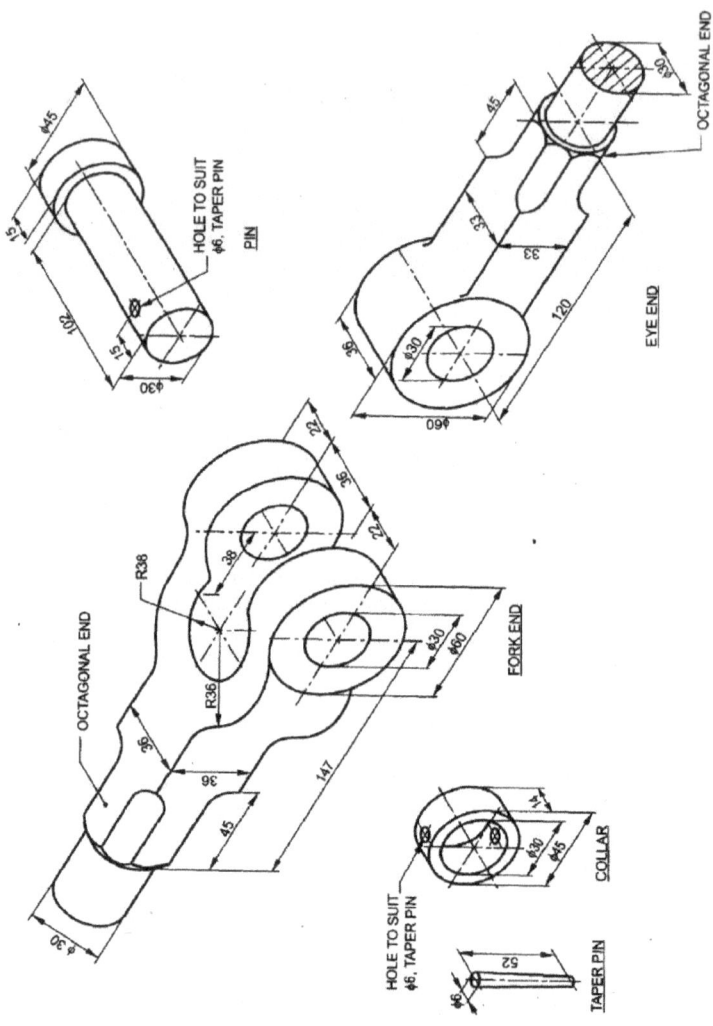

Fig.1 Knuckle Joint

ME 6611 CAD/CAM Laboratory
Department of Mechanical Engineering

SOLID MODELING AND ASSEMBLY OF KNUCKLE JOINT

OBJECTIVE:

To draw the components of Knuckle Joint and to assemble them using Solid works software.

SOFTWARE:

Solid Works 2008.

DESCRIPTION ABOUT STUFFING BOX:

Knuckle joint is a type of mechanical joint used in structures, to connect two intersecting cylindrical rods, whose axes lie on the same plane. It permits some angular movement between the cylindrical rods (in their plane). It is specially designed to withstand tensile loads. Coaxial holes are provided in the fork end, eye end and collar. The fork end and the eye end are held together in position by means of a knuckle pin. The knuckle pin is held in its position with the help of a collar and a taper pin.

The assembled view of a knuckle joint is shown in the image below. Both the fork end and the eye end are capable of rotating in their planes about the axis of the knuckle pin.

PROCEDURE:

1. Identify various parts to be created.

2. First enter into part environment and create the main part and create the main part of the assembly.

3. First identify whether the main part or the first to be created by protrusion or by revolution.

4. Select the sketch tool and then select the coincidental plane option and select any one of the standard 3 planes (i.e. front, right &top).

5. Create the cross-section profile as a closed one using the 2D commands available after completing the sketch, click open or return button and then click finish button.

6. For creating other parts, select sketch both parallel plane option or plane by 3points option and then select the required plane.

7. Construct the full cross section for portion and construct the half of the cross section and an axis line for revolution.

8. Do the protrusions by using protrusion command and the revolution by revolved protrusion command.

9. For constructing holes and cut-out, used hole command and cut-out command.

10. If we use hole command, change the diameter of the hole by using modify menu, resize hole option.

11. Use revolved cut-out command whenever needed.

12. Use the distance between option to maintain accurate distance between one edge and other edge or between one edge and to center the hole.

13. After constructing each part save it as a separate part file with extrusion* par.

14. Enter into assembly environment.

15. Assembly the various parts construct parts construct using the various assembly constrains available (planer, design, mate, axial align, connect etc.).

16. After finishing assembly, check whether the various parts have been connected properly or not by rotating the view.

17. Save the assembly.

ASSEMBLY 3D DRAWING:

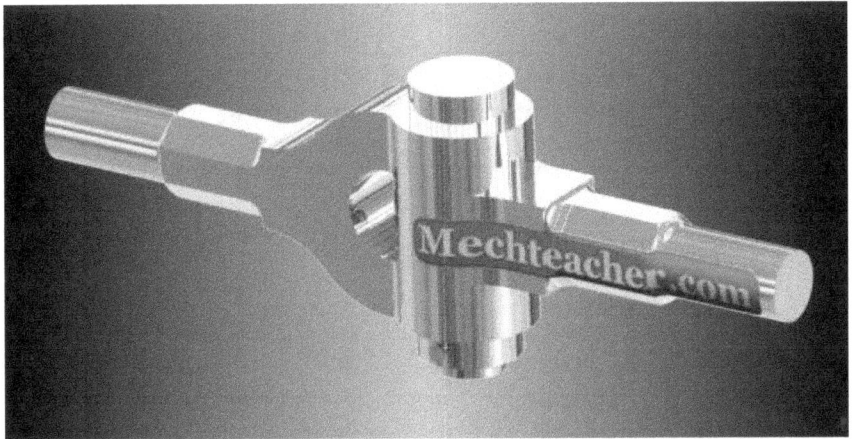

RESULTS & DISCUSSION:

_____

_____

_____

_____

_____

REPORT EVALUATION

S.NO	DESCRIPTION	WEIGHTAGE	MARK AWARDED
1	Part modeling	30	
2	Assembly & Drafting	30	
3	Results & Discussion	20	
4	Student Performance	10	
6	Viva-voce	10	
	Total	100	

Signature of Course Coordinator

EXPERIMENT 7

SOLID MODELING AND ASSEMBLY OF LATHE TAILSTOCK

Name: _____

Register Number: _____

Report Submitted on: _____

Important Instructions:

1. Lab reports must be submitted on the next working day of the lab class. Late submissions will entitle reduction in marks allotted for reports.

2. Plots must be glued neatly on the space provided.

3. All observations and programs should be hand written. Avoid attaching printouts.

4. Individual reports must be submitted for all the lab experiments.

5. Plagiarism will be treated very seriously.

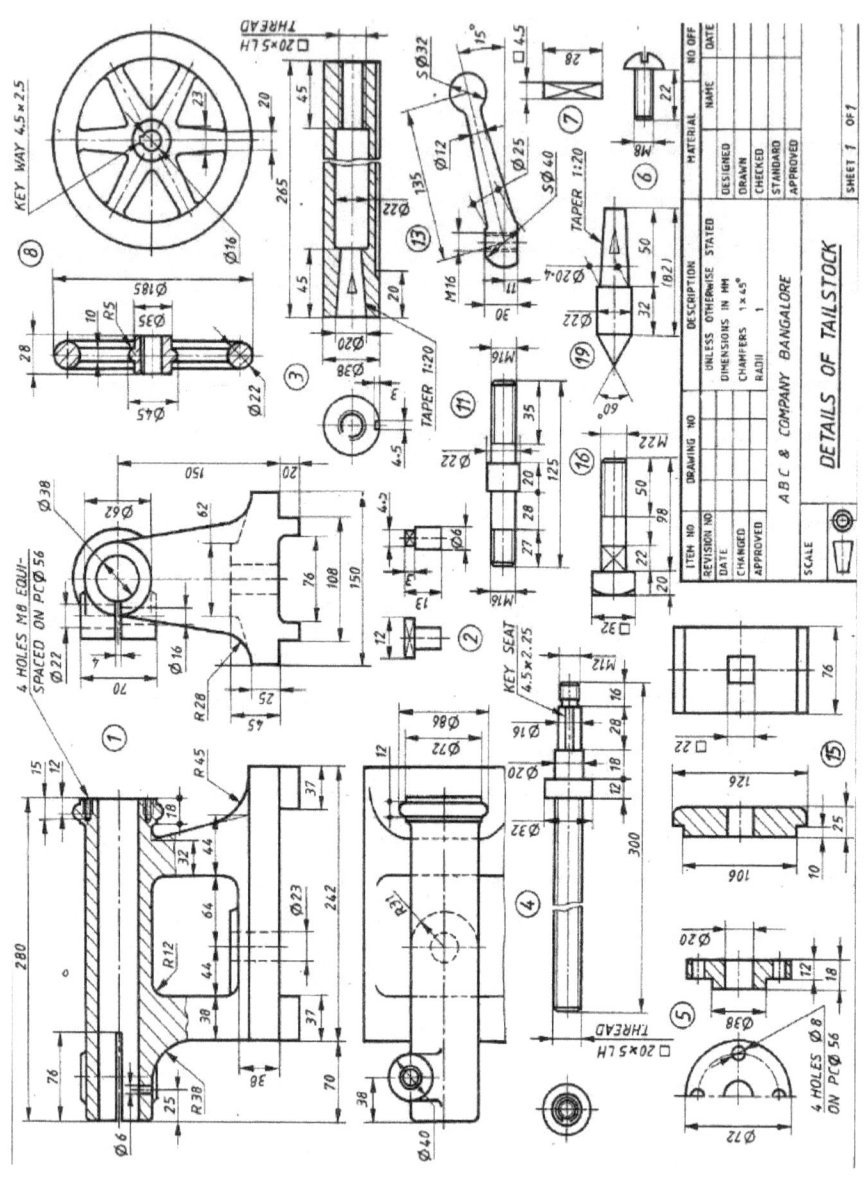

Fig.1 Lathe Tailstock

SOLID MODELING AND ASSEMBLY OF LATHE TAILSTOCK

OBJECTIVE:

To draw the components of lathe tailstock and to assemble them using Solid works software.

SOFTWARE:

Solid Works 2008

DESCRIPTION ABOUT LATHE TAILSTOCK:

A **tailstock**, also known as a **foot stock**, is a device often used as part of an engineering lathe, wood-turning lathe, or used in conjunction with a rotary table on a milling machine.

It is usually used to apply support to the longitudinal rotary axis of a workpiece being machined. A lathe center is mounted in the tailstock, and inserted against the sides of a hole in the center of the workpiece. A tailstock has a Dead Center, while Head stock has a Live Center. A Tailstock is particularly useful when the workpiece is relatively long and slender. Failing to use a tailstock can cause "chatter," where the workpiece bends excessively while being cut.

It is also used on a lathe to hold drilling or reaming tools for machining a hole in the work piece. Unlike drilling with a drill press or a milling machine, the tool is stationary while the workpiece rotates. Holes can only be cut along the axis that the workpiece is set to spin.

Usually, the entire tailstock is moved to the approximate position that it will be needed by manually sliding it along its ways. There, it is locked in place and the tool mounted to it is moved with a lead screw to the exact position where it is needed. When a cutting tool such as a drill bit or reamer is used, the feed is done with this lead screw. The tailstock quill or extendible portion usually has a Morse taper mount in the end of it to secure the   drill   or   reamer.

PROCEDURE:

1. Identify various parts to be created.
2. First enter into part environment and create the main part and create the main part of the assembly.
3. First identify whether the main part or the first to be created by protrusion or by revolution.
4. Select the sketch tool and then select the coincidental plane option and select any one of the standard 3 planes (i.e. front, right & top).
5. Create the cross-section profile as a closed one using the 2D commands available after completing the sketch, click open or return button and then click finish button.
6. For creating other parts, select sketch both parallel plane option or plane by 3points option and then select the required plane.
7. Construct the full cross section for portion and construct the half of the cross section and an axis line for revolution.
8. Do the protrusions by using protrusion command and the revolution by revolved protrusion command.
9. For constructing holes and cut out, used hole command and cut out command.
10. If we use hole command, change the diameter of the hole by using modify menu, resize hole option.
11. Use revolved cut out command whenever needed.
12. Use the distance between option to maintain accurate distance between one edge and other edge or between one edge and to center the hole.
13. After constructing each part save it as a separate part file with extrusion par.
14. Enter into assembly environment.
15. Assembly the various parts construct parts construct using the various assembly constrains available (planer, design, mate, axial align, connect etc.).
16. After finishing assembly, check whether the various parts have been connected properly or not by rotating the view.
17. Save the assembly.

ASSEMBLY 3D DRAWING:

RESULTS & DISCUSSION:

_____

_____

_____

_____

_____

VIVA QUESTIONS:

1. What is lathe tail stock?
2. What are the types of stock in lathe?
3. What is the use of tail stock?
4. What is specification of tail stock?
5. What are advantages of tail stock?

REPORT EVALUATION

S.NO	DESCRIPTION	WEIGHTAGE	MARK AWARDED
1	Part modeling	30	
2	Assembly & Drafting	30	
3	Results & Discussion	20	
4	Student Performance	10	
6	Viva-voce	10	
	Total	100	

Signature of Course Instructor

EXPERIMENT 8

SOLID MODELING AND ASSEMBLY OF CONNECTING ROD

Name: _____

Register Number: _____

Report Submitted on: _____

Important Instructions:

1. Lab reports must be submitted on the next working day of the lab class. Late submissions will entitle reduction in marks allotted for reports.

2. Plots must be glued neatly on the space provided.

3. All observations and programs should be hand written. Avoid attaching printouts.

4. Individual reports must be submitted for all the lab experiments.

5. Plagiarism will be treated very seriously.

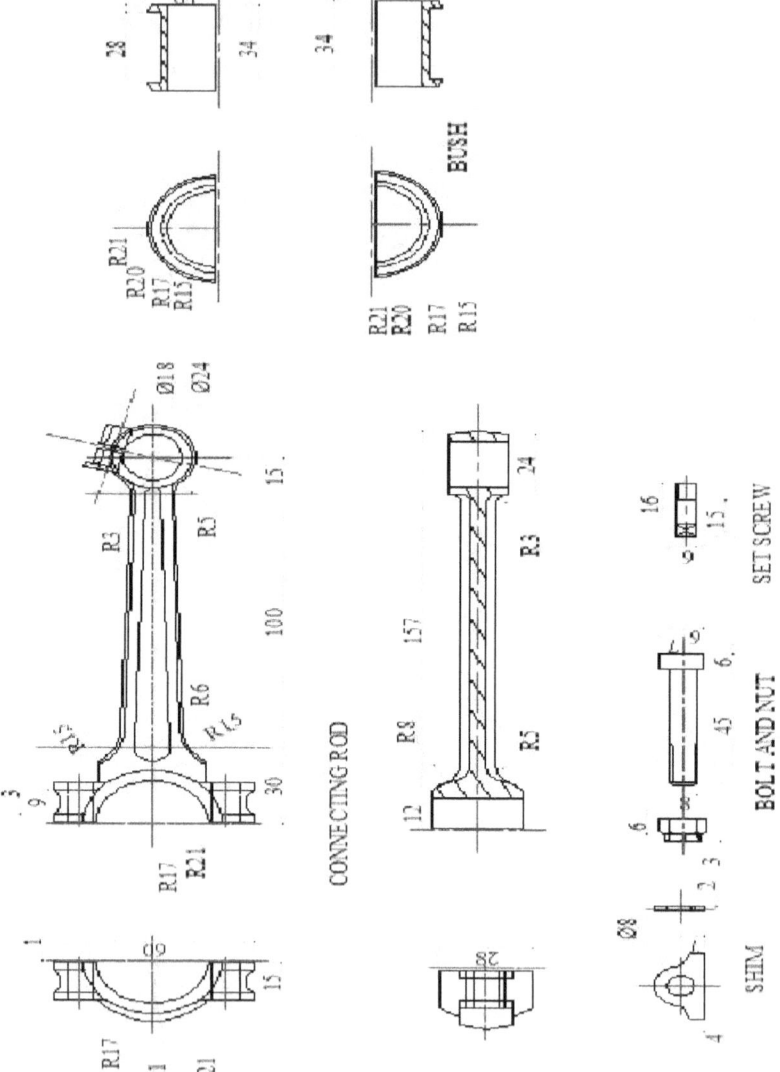

ALL DIMENSIONS ARE IN "MM"

BUSH

CONNECTING ROD

SET SCREW

BOLT AND NUT

SHIM

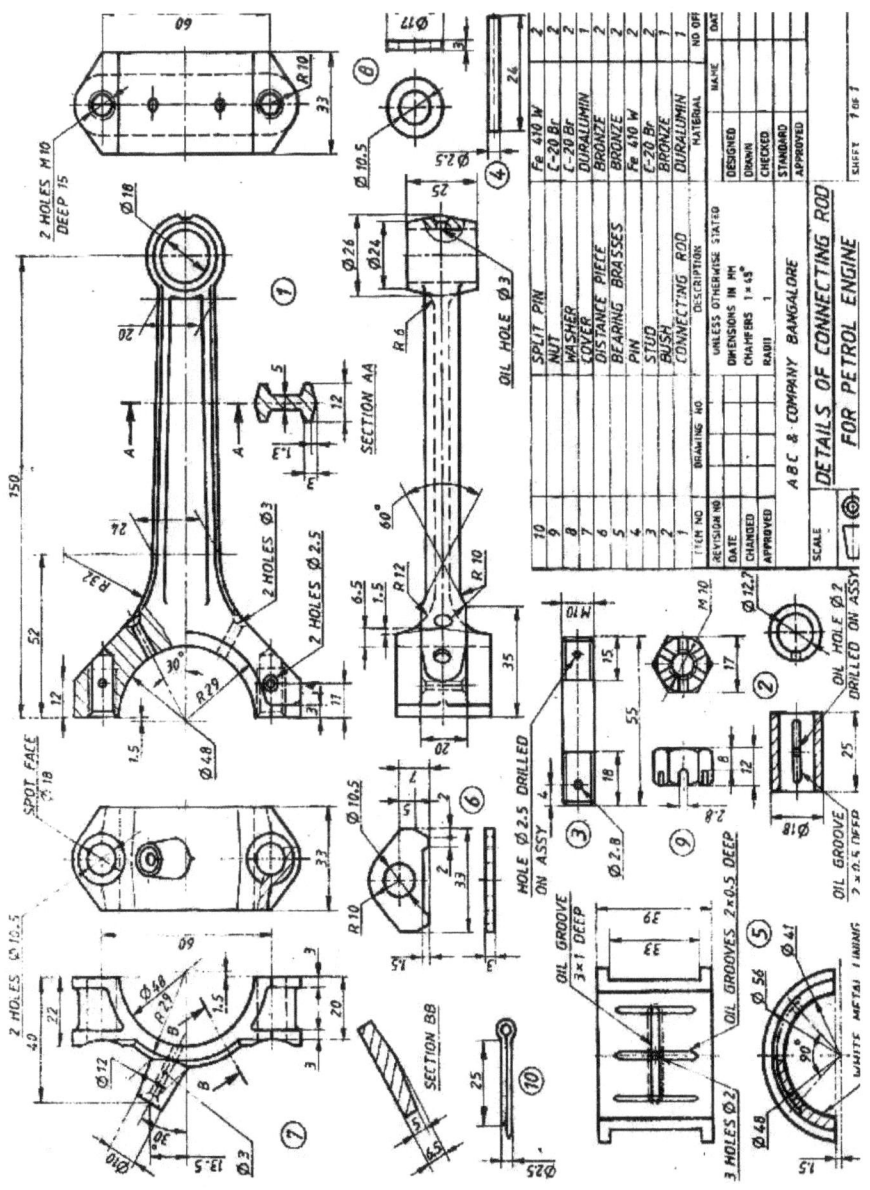

Fig.1 Connecting rod

SOLID MODELING AND ASSEMBLY OF CONNECTING ROD

OBJECTIVE:

To draw the components of connecting rod and to assemble them using Solid works software.

SOFTWARE:

Solid Works 2008

DESCRIPTION ABOUT CONNECTING ROD:

A connecting rod is a shaft which connects a piston to a crank or crankshaft in a reciprocating engine. Together with the crank, it forms a simple mechanism that converts reciprocating motion into rotating motion.

A connecting rod may also convert rotating motion into reciprocating motion, its original use. Earlier mechanisms, such as the chain, could only impart pulling motion. Being rigid, a connecting rod may transmit either push or pull, allowing the rod to rotate the crank through both halves of a revolution. In a few two-stroke engines the connecting rod is only required to push.

Today, the connecting rod is best known through its use in internal combustion piston engines, such as automobile engines. These are of a distinctly different design from earlier forms of connecting rod used in steam engines and steam locomotives.

PROCEDURE:

1. Identify various parts to be created.
2. First enter into part environment and create the main part and create the main part of the assembly.
3. First identify whether the main part or the first to be created by protrusion or by revolution.
4. Select the sketch tool and then select the coincidental plane option and select any one of the standard 3 planes (i.e. front, right & top).
5. Create the cross-section profile as a closed one using the 2D commands available after completing the sketch, click open or return button and then click finish button.
6. For creating other parts, select sketch both parallel plane option or plane by 3points option and then select the required plane.
7. Construct the full cross section for portion and construct the half of the cross section and an axis line for revolution.
8. Do the protrusions by using protrusion command and the revolution by revolved protrusion command.
9. For constructing holes and cut out, used hole command and cut out command.
10. If we use hole command, change the diameter of the hole by using modify menu, resize hole option.
11. Use revolved cut out command whenever needed.
12. Use the distance between option to maintain accurate distance between one edge and other edge or between one edge and to center the hole.
13. After constructing each part save it as a separate part file with extrusion par.
14. Enter into assembly environment.
15. Assembly the various parts construct parts construct using the various assembly constrains available (planer, design, mate, axial align, connect etc.).
16. After finishing assembly, check whether the various parts have been connected properly or not by rotating the view.
17. Save the assembly.

ASSEMBLY 3D DRAWING:

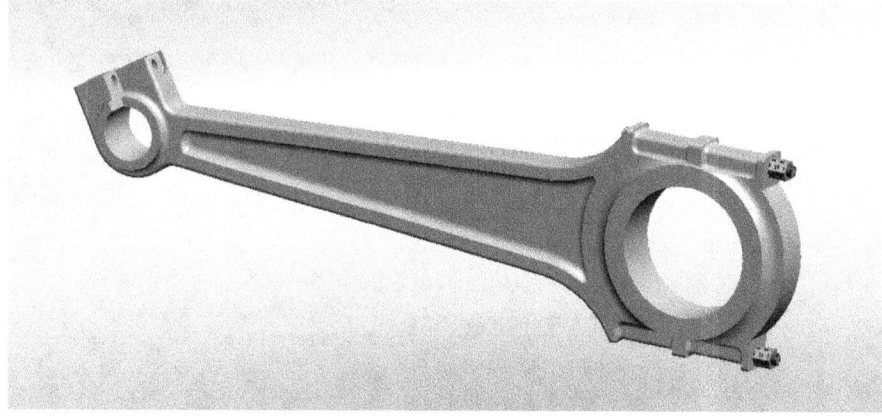

RESULTS & DISCUSSION:

_____

_____

_____

_____

_____

VIVA QUESTIONS:

1. What is connecting rod?
2. What is the use of connecting rod?
3. What is specification of connecting rod?
4. What are advantages of connecting rod?

REPORT EVALUATION

S.NO	DESCRIPTION	WEIGHTAGE	MARK AWARDED
1	Part modeling	30	
2	Assembly & Drafting	30	
3	Results & Discussion	20	
4	Student Performance	10	
6	Viva-voce	10	
Total		100	

Signature of Course Coordinator

2 DIMENSIONAL DRAWING OF LEVER SAFETY VALVE

Name: _____

Register Number: _____

Report Submitted on: _____

Important Instructions:

1. Lab reports must be submitted on the next working day of the lab class. Late submissions will entitle reduction in marks allotted for reports.

2. Plots must be glued neatly on the space provided.

3. All observations and programs should be hand written. Avoid attaching printouts.

4. Individual reports must be submitted for all the lab experiments.

5. Plagiarism will be treated very seriously.

2 DIMENSIONAL DRAWING OF LEVER SAFETY VALVE

OBJECTIVE:

To draw the components of safety valve in a two dimensional.

2D DRAWING:

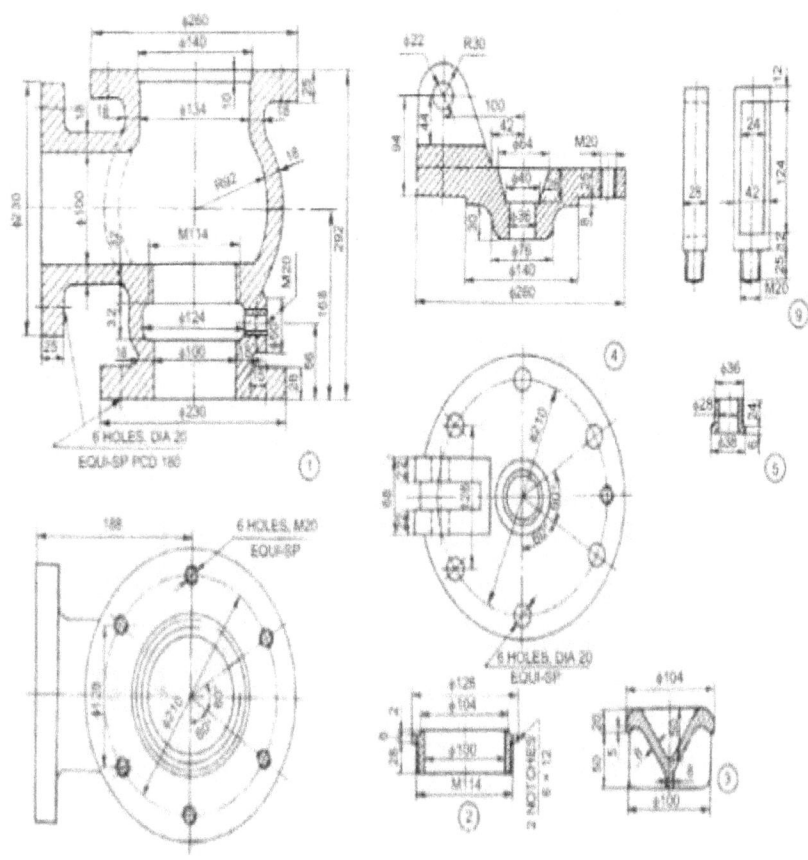

Fig1. Lever Safety valve

ASSEMBLY 3D DRAWING:

RESULTS & DISCUSSION:

_____

_____

_____

_____

_____

REPORT EVALUATION

S.NO	DESCRIPTION	WEIGHTAGE	MARK AWARDED
1	Part modeling	30	
2	Assembly & Drafting	30	
3	Results & Discussion	20	
4	Student Performance	10	
6	Viva-voce	10	
Total		100	

Signature of Course Coordinator

EXPERIMENT 10

2 DIMENSIONAL DRAWING OF NON RETURN VALVE

Name: _____

Register Number: _____

Report Submitted on: _____

Important Instructions:

1. Lab reports must be submitted on the next working day of the lab class. Late submissions will entitle reduction in marks allotted for reports.

2. Plots must be glued neatly on the space provided.

3. All observations and programs should be hand written. Avoid attaching printouts.

4. Individual reports must be submitted for all the lab experiments.

5. Plagiarism will be treated very seriously.

2 DIMENSIONAL DRAWING OF NON RETURN VALVE

OBJECTIVE:

To draw the components of safety valve in a two dimensional.

2D DRAWING:

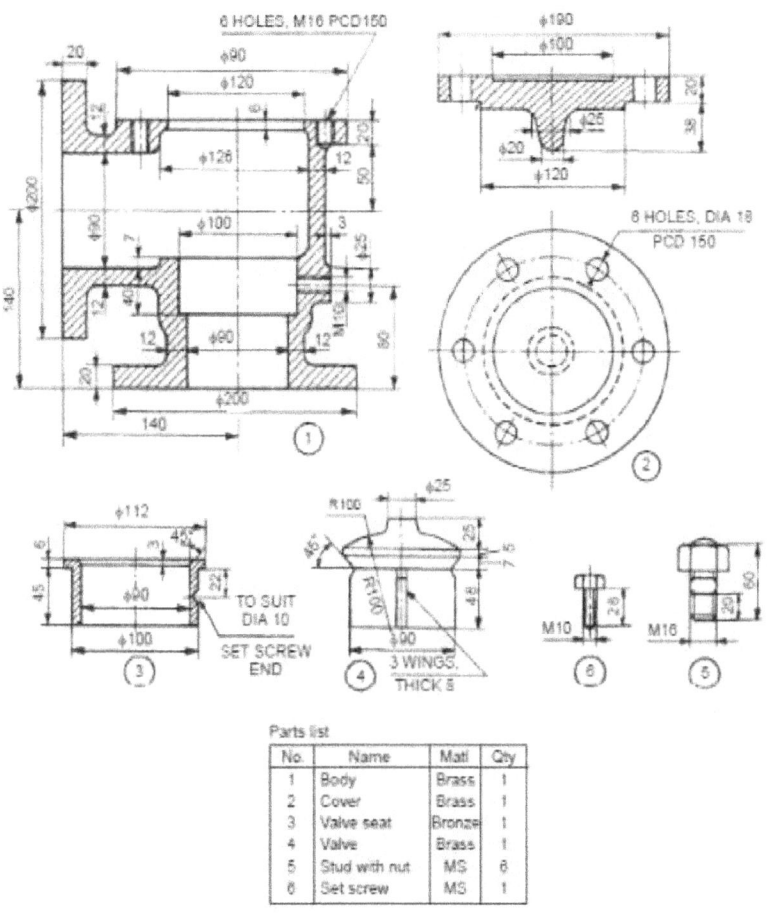

Fig1. Non return valve

ASSEMBLY 3D DRAWING:

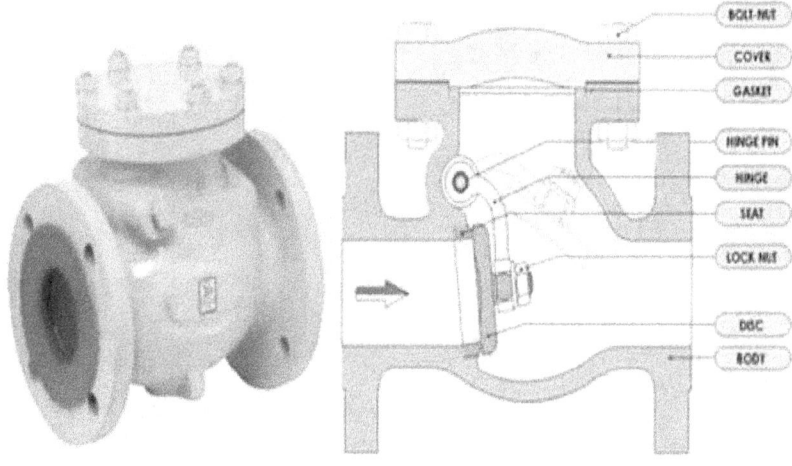

BOLT-NUT
COVER
GASKET
HINGE PIN
HINGE
SEAT
LOCK NUT
DISC
BODY

ME 6611 CAD/CAM Laboratory
Department of Mechanical Engineering

RESULTS & DISCUSSION:

_____

_____

_____

_____

_____

REPORT EVALUATION

S.NO	DESCRIPTION	WEIGHTAGE	MARK AWARDED
1	Part modeling	30	
2	Assembly & Drafting	30	
3	Results & Discussion	20	
4	Student Performance	10	
6	Viva-voce	10	
	Total	100	

Signature of Course Coordinator

EXPERIMENT 1

TURNING USING CIRCULAR INTERPOLATIONS (G03, G04)

Name: _____

Register Number: _____

Report Submitted on: _____

Important Instructions:

1. Lab reports must be submitted on the next working day of the lab class. Late submissions will entitle reduction in marks allotted for reports.

2. Plots must be glued neatly on the space provided.

3. All observations and programs should be hand written. Avoid attaching printouts.

4. Individual reports must be submitted for all the lab experiments.

5. Plagiarism will be treated very seriously.

TURNING USING CIRCULAR INTERPOLATIONS (G03, G04)

OBJECTIVE

To write the part programming and simulation them to the given lathe job.

PROCEDURE

1. To write the program for given job.

2. To type G and M CODES.

3. To give the tool size and stock dimensions.

4. Finally to run the machine to the operation.

PROGRAM

```
O0001;
N1 G28 U0 W0;
G50 S1000;
G96 S200 M03;
M06 T0101 M08;
G00 X36.0 Z1.0;
G71 U0.5 R3;
G71 P2 Q3 U0.15 W0.15 F0.2;
N2 G01 X10.0 F0.2;
G01 Z-20.0 F0.2;
G02 X20.0 Z-30.0 R10.0 F0.2;
G03 X35.0 Z-37.50 R7.5 F0.2;
G01 Z-60.0 F0.2;
N3 G00 X37.0;
G28 U0 W0;
M05;
M09;
G28 U0 W0;
M06 T0202;
G97 S600 M03;
G00 X36.0 Z1.0;
N4 G70 P2 Q3 F0.15;
G28 U0 W0;
M05;
M09;
M30;
```

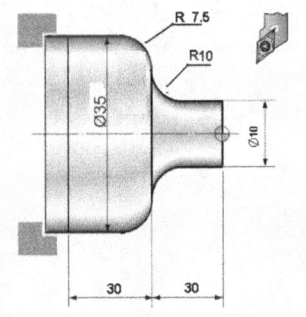

RESULTS & DISCUSSION:

_____

_____

_____

REPORT EVALUATION

S.NO	DESCRIPTION	WEIGHTAGE	MARK AWARDED
1	CAM Programming	30	
2	Machining	30	
3	Results & Discussion	20	
4	Student Performance	10	
6	Viva-voce	10	
	Total	100	

Signature of Course Coordinator

EXPERIMENT 2

DRILLING USING DRILLING CYCLE (G73)

Name: _____

Register Number: _____

Report Submitted on: _____

Important Instructions:

1. Lab reports must be submitted on the next working day of the lab class. Late submissions will entitle reduction in marks allotted for reports.

2. Plots must be glued neatly on the space provided.

3. All observations and programs should be hand written. Avoid attaching printouts.

4. Individual reports must be submitted for all the lab experiments.

5. Plagiarism will be treated very seriously.

DRILLING USING DRILLING CYCLE (G73)

OBJECTIVE

To write the part programming and simulation them to the given lathe job.

PROCEDURE

1. To write the program for given job.

2. To type G and M CODES.

3. To give the tool size and stock dimensions.

4. Finally to run the machine to the operation.

PROGRAM

```
O0002;
N1 G28 U0 W0;
G50 S1000;
G97 S800 M03;
M06 T0101 M08;
G00 X0.0 Z1.0;
G74 R2;
G74 Z-30.0 Q1000 F0.2;
N2 G00 Z5.0;
G28 U0 W0;
M05;
M09;
M30;
```

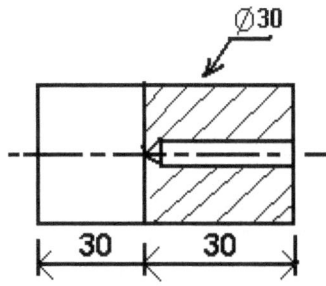

RESULTS & DISCUSSION:

_____

_____

_____

REPORT EVALUATION

S.NO	DESCRIPTION	WEIGHTAGE	MARK AWARDED
1	CAM Programming	30	
2	Machining	30	
3	Results & Discussion	20	
4	Student Performance	10	
6	Viva-voce	10	
Total		100	

Signature of Course Coordinator

EXPERIMENT 3

GROOVING USING GROOVING CYCLE (G75)

Name: _____

Register Number: _____

Report Submitted on: _____

Important Instructions:

1. Lab reports must be submitted on the next working day of the lab class. Late submissions will entitle reduction in marks allotted for reports.

2. Plots must be glued neatly on the space provided.

3. All observations and programs should be hand written. Avoid attaching printouts.

4. Individual reports must be submitted for all the lab experiments.

5. Plagiarism will be treated very seriously.

GROOVING USING GROOVING CYCLE (G75)

OBJECTIVE

To write the part programming and simulation them to the given lathe job.

PROCEDURE

1. To write the program for given job.

2. To type G and M CODES.

3. To give the tool size and stock dimensions.

4. Finally to run the machine to the operation.

PROGRAM

O0003;
N1 G28 U0 W0;
G50 S1000;
G97 S800 M03;
M06 T0101 M08;
G00 X32.0 Z1.0;
G75 R2;
G75 X10 Z-30.0 P500 Q1000 F0.2;
N2 G00 X32.0 Z1.0;
G28 U0 W0;
M05;
M09;
M30;

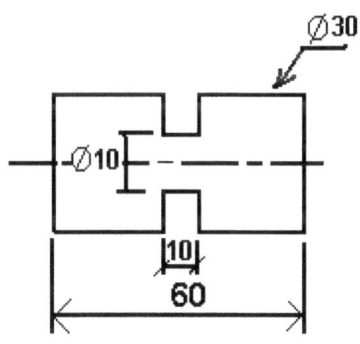

RESULTS & DISCUSSION:

_____

_____

_____

REPORT EVALUATION

S.NO	DESCRIPTION	WEIGHTAGE	MARK AWARDED
1	CAM Programming	30	
2	Machining	30	
3	Results & Discussion	20	
4	Student Performance	10	
6	Viva-voce	10	
Total		100	

Signature of Course Coordinator

EXPERIMENT 4

THREADING USING THREADING CYCLE (G76)

Name: _____

Register Number: _____

Report Submitted on: _____

Important Instructions:

1. Lab reports must be submitted on the next working day of the lab class. Late submissions will entitle reduction in marks allotted for reports.

2. Plots must be glued neatly on the space provided.

3. All observations and programs should be hand written. Avoid attaching printouts.

4. Individual reports must be submitted for all the lab experiments.

5. Plagiarism will be treated very seriously.

THREADING USING THREADING CYCLE (G76)

OBJECTIVE

To write the part programming and simulation them to the given lathe job

PROCEDURE

1. To write the program for given job.

2. To type G and M CODES.

3. To give the tool size and stock dimensions.

4. Finally to run the machine to the operation.

PROGRAM

```
O0004;
N1 G28 U0 W0;
G50 S1000;
G97 S800 M03;
M06 T0101 M08;
G00 X32.0 Z1.0;
G76 P021060 QI00 R100;
G76 X8.5 Z-20 P900 Q200 FI.5;
G00 X32.0 Z1.0;
G76 P021060 QI00 R100;
G76 X18.5 Z-52 P900 Q200 FI .5;
G00 X32.0 Z1.0;
G28 U0 W0;
M05;
M09;
M30;
```

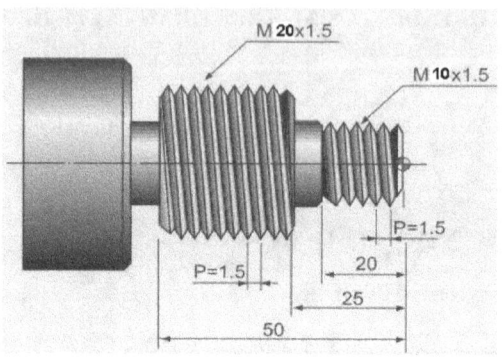

RESULTS & DISCUSSION:

_____

_____

_____

REPORT EVALUATION

S.NO	DESCRIPTION	WEIGHTAGE	MARK AWARDED
1	CAM Programming	30	
2	Machining	30	
3	Results & Discussion	20	
4	Student Performance	10	
6	Viva-voce	10	
Total		100	

Signature of Course Coordinator

EXPERIMENT 5

FACE MILLING

Name: _____

Register Number: _____

Report Submitted on: _____

Important Instructions:

1. Lab reports must be submitted on the next working day of the lab class. Late submissions will entitle reduction in marks allotted for reports.

2. Plots must be glued neatly on the space provided.

3. All observations and programs should be hand written. Avoid attaching printouts.

4. Individual reports must be submitted for all the lab experiments.

5. Plagiarism will be treated very seriously.

FACE MILLING

OBJECTIVE

To write the part programming and simulation them to the given milling job.

PROCEDURE

1. To write the program for given job.

2. To type G and M CODES.

3. To give the tool size and stock dimensions.

4. Finally to run the machine to the operation.

PROGRAM

```
O1001;
N1 G21 G94;
N2 G91 G28 X0 Y0 Z0;
N3 T01 M06;
N4 G90 G54 G00 X25.0 Y25.0;
N5 G43 H01 Z100.0;
N6 M03 S1000 Z20;
N7 G01 Z-5.0 F400;
N8 Y175.0;
N9 X175.0;
N10 Y25.0;
N11 X25.0;
N12 G00 Z100.0;
N13 G91 G28 X0 Y0 Z0;
N14 M30;
```

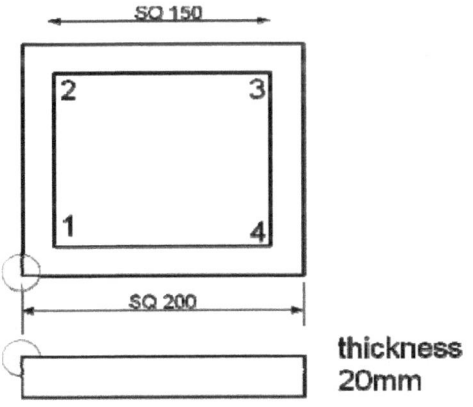

Making 150 mm* 150 mm square for a depth of 5mm in a given Billet

RESULTS & DISCUSSION:

_____

_____

_____

REPORT EVALUATION

S.NO	DESCRIPTION	WEIGHTAGE	MARK AWARDED
1	CAM Programming	30	
2	Machining	30	
3	Results & Discussion	20	
4	Student Performance	10	
6	Viva-voce	10	
	Total	100	

Signature of Course Coordinator

EXPERIMENT 6

DRILLING FOR SIZE 100*100*20 Ø8, FIVE HOLES

Name: _____

Register Number: _____

Report Submitted on: _____

Important Instructions:

1. Lab reports must be submitted on the next working day of the lab class. Late submissions will entitle reduction in marks allotted for reports.

2. Plots must be glued neatly on the space provided.

3. All observations and programs should be hand written. Avoid attaching printouts.

4. Individual reports must be submitted for all the lab experiments.

5. Plagiarism will be treated very seriously.

DRILLING FOR SIZE 100*100*20 Ø8, FIVE HOLES

OBJECTIVE

To write the part programming and simulation them to the given milling job.

PROCEDURE

1. To write the program for given job.

2. To type G and M CODES.

3. To give the tool size and stock dimensions.

4. Finally to run the machine to the operation.

PROGRAM

```
O1002;
N1 G91 G28 X0 Y0 Z0;
N2 T01 M06;
N3 G21 G94;
N4 G17 G90 G54 G00 X20 Y20;
N5 G43 H01 Z100;
N6 M03 S1000 Z20;
N7 G99 G81 Z-4.0 R5 F300;
N8 X80;
N9 Y80;
N10 X20;
N11 X50 Y50;
N12 G80;
N13 G00 Z100;
N14 G91 G28 X0 Y0 Z0;
N15 M05;
N16 M30;
```

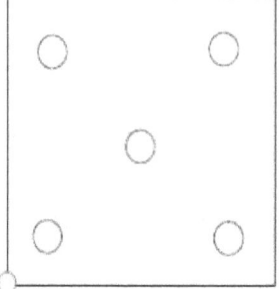

```
BLANK SIZE 100*100*20
DIA. 8, FIVE HOLES
HOLE1    (20,20)
HOLE2    (20,80)
HOLE3    (80,80)
HOLE4    (80,20)
HOLE5    (50,50)
```

RESULTS & DISCUSSION:

_____

_____

_____

REPORT EVALUATION

S.NO	DESCRIPTION	WEIGHTAGE	MARK AWARDED
1	CAM Programming	30	
2	Machining	30	
3	Results & Discussion	20	
4	Student Performance	10	
6	Viva-voce	10	
Total		100	

Signature of Course Coordinator

DRILLING & TAPPING FOR SIZE 100*100*20 Ø8, FIVE HOLES

Name: _____

Register Number: _____

Report Submitted on: _____

Important Instructions:

1. Lab reports must be submitted on the next working day of the lab class. Late submissions will entitle reduction in marks allotted for reports.

2. Plots must be glued neatly on the space provided.

3. All observations and programs should be hand written. Avoid attaching printouts.

4. Individual reports must be submitted for all the lab experiments.

5. Plagiarism will be treated very seriously.

DRILLING & TAPPING FOR SIZE 100*100*20 Ø8, FIVE HOLES

OBJECTIVE

   To write the part programming and simulation them to the given milling job.

PROCEDURE

   1. To write the program for given job.

   2. To type G and M CODES.

   3. To give the tool size and stock dimensions.

   4. Finally to run the machine to the operation.

PROGRAM

   O1002;
   N10 G91 G28 X0 Y0 Z0;
   N20 T02 M06;
   N30 G21 G99;
   N40 G17 G90 G54 G00 X20 Y20;
   N50 G43 H01 Z100;
   N60 M03 S1000 Z20;
   N70 G99 G84 Z-15.0 R5 P1000 F300; (FEED = RPM* PITCH)
   N80 X80;
   N90 Y80;
   N100 X20;
   N110 X50 Y50;
   N120 G80 G00 Z100;
   N130 G91 G28 X0 Y0 Z0;
   N140 M05;
   N150 M30;

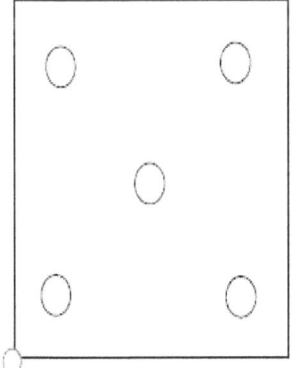

BLANK SIZE 100*100*20

DIA. 8, FIVE HOLES

HOLE1	(20,20)
HOLE2	(20,80)
HOLE3	(80,80)
HOLE4	(80,20)
HOLE5	(50,50)

RESULTS & DISCUSSION:

_____

_____

_____

REPORT EVALUATION

S.NO	DESCRIPTION	WEIGHTAGE	MARK AWARDED
1	CAM Programming	30	
2	Machining	30	
3	Results & Discussion	20	
4	Student Performance	10	
6	Viva-voce	10	
	Total	100	

Signature of Course Coordinator

VARIOUS CYCLES USED IN THE CNC LATHE

A. FACING CYCLE

Format: G72 W(1) --- R---;
               G72 P---Q---U---W(2) ---F---;

W(1)	-	Depth of cut in z axis
R	-	Relief amount in x axis
P	-	Sequence number of the first block of the program
Q	-	Sequence number of the last block of the program
U		- Finish allowance in x axis
W(2)	-	Finish allowance in z axis
F	-	Feed rate

B. TURNING

Format: G71 W(1) --- R---;
               G71 P---Q---U---W (2) ---F---;

W(1)	-	Depth of cut in x axis
R	-	Relief amount in y axis & x axis
P	-	Sequence number of the first block of the program
Q	-	Sequence number of the last block of the program
U	-	Finish allowance in x axis
W(2)	-	Finish allowance in z axis
F	-	Feed rate

C. THREAD CUTTING

1. G76: MULTIPLE THREAD CYCLE

Format: G76P _ _ _Q(1) _ R _;
                G76X_Z_P_Q(2)_R_F_;

P020000 (eg.) -   02 Number of ideal pass (after cutting   thread)
                                00 Retract angle
                                00 Thread angle

Q (1)	-	Radial depth of cut in regular pass
R	-	Finishing allowance in the last pass
X	-	Thread minor dia value (in case of ext. thread) (or) Thread major dia. value (in case of int. thread)
Z	-	Thread length
P	-	Thread depth (one side depth)
Q (2)	-	Radial depth of cut in the first pass
R	-	Taper value
F	-	Pitch of the thread

D. G74 DRILLING CYCLE

Format: G74R--;
                 G74Z---Q---F---;

R   -   Return amount in z axis
Z   -   Drill hole depth
Q   -   Incremental depth of cut in z axis
F   -   Feed rate

F. G75 GROOVING CYCLE

Format: G75R---;
               G75X---Z---P---Q---F---;

R   -   Relief amount in x axis
X   -   Groove diameter
Z   -   Groove length /Groove end point
P   -   Depth of cut in x axis
Q   -   Incremental depth of cut in z axis
F   -   Feed rate

G70 FINISHING CYCLE:

A G70 causes a range of blocks to be executed/ then control passes to the block after the G70. This will be used after the completion of the roughing cycle. The P and Q values specify the "N" block numbers at the start and end of the profile.

Example: G70P10Q20

P- First block of cycle
Q- Last block of cycle

G71 MULTIPLE TURNING CYCLE:

A G71 causes the profile to be roughed out by turning. Control passes on to after the last block of the profile. Two G71 blocks are needed to specify all the values.

Example: i) G71U2.0R1.5
               ii) G71P10Q20U0.1W0.1F25

i) G71U2.0R1.0

U □ Depth of cut in mm
R □ Retraction (or) Retardation amount in mm

ii) G71P10Q20U0.1W0.1F25

P - Starting block number (i.e.) first block of the cycle.
Q - End block number
U - Finishing allowance along X axis in mm
W - Finishing allowance along Z axis in mm
F - Feed rate

DESCRIPTION OF G CODES

G00 FAST TRAVERSE

A G00 causes linear motion to the given position at the maximum feed rate from the current position that is predefined in the option file.

Examples: G00X0.0Y0.0

G01 LINEAR INTERPOLATION:

A G01 causes linear motion to the position at the last specified feed rate from the current position. The feed rate for the linear motion should be mentioned in the part program.

Example: G01X30.0Y10.0F100.0

G02 CIRCULAR INTERPOLATION (CW)

A G02 causes a clockwise arc to the specified position.

Example: G02X30.0Y20.0R10.0

G03 CIRCULAR INTERPOLATION (CCW)

A G02 causes a counter clockwise arc to the specified position.

Example: G03X30Y20R20

G21 METRIC:

A G21 cause positions to be interpreted as being in metric units (mm). This can only be at the main program. By default metric units will be taken for programming.

G28 GOTO REFERENCE POINT:

A G28 causes a fast traverse to the specified position and then to the machine datum.

Example: G28U0.0W0.0

G90 ABSOLUTE MOVEMENT:

All future movement will be absolute until overridden by a G91 instruction. This is the default setting.

Example: G90
             G01X30Y0
The position becomes (X30, Y0), irrespective of the previous position.

G91 INCREMENTAL MOVEMENT:

All future movement will be incremental (i.e. relative to the current position of the tool) until overridden by a G90 instruction.

> Example: G90
>   G01X15Y10
> G91
> G01X2
> The position becomes (X17, Y10).

PREPARATORY FUNCTION

(G -CODES)

- G00- Fast transverse
- G01- Linear interpolation
- G02- Circular interpolation (c.w)
- G03- Circular interpolation (c.c.w)
- G04-Dwell
- G20-Imporial (input in inches)
- G21- Metric (input in mm)
- G28- Go to reference
- G40- Cutter compensation cancel
- G41- Cutter compensation right
- G42-Cutter compensation left
- G50- Co-ordinate setting
- G70-Finishing cycle
- G71- Stock removal in turning
- G72- Multiple facing
- G73-Pattern repeating
- G74- drilling
- G76- Multiple thread
- G81- Drilling cycle
- G90-Turning cycle
- G94- Facing cycle
- G96- Constant surface
- G97- Variable surface
- G98- Feed per minute
- G99- Feed per revolution

MISCELLANEOUS FUNCTION

(M - CODES)

- M00- Program stop
- M02- Optional stop
- M03- Program end
- M04- Spindle forward
- M05- Spindle stop
- M06- Tool change
- M08- Coolant on
- M09- Coolant off
- M10- Vice open
- M11- Vice close
- M62- Output 1ON
- M63- Output 2ON
- M64- Output1OFF
- M65- Output 2OFF
- M60- Wait input 1ON
- M67- Wait input 1OFF
- M76- Wait input 2OFF
- M77-Sub program call
- M98-Sub program exit
- M99- Sub program exit
- M30- Program and rewind

www.ingramcontent.com/pod-product-compliance
Lightning Source LLC
Chambersburg PA
CBHW071222220526
45468CB00002B/704